Rocks and Landforms of the Pembrokeshire Coast National Park

Dyfed Ellis-Gruffydd

Information Booklet No. 1
Published by

Greencroft Books

for the Pembrokeshire Coast
National Park Committee

1977

Greencroft Books

Published by Brian S. John, Greencroft Books, Trefelin, Newport, Dyfed, for the Pembrokeshire Coast National Park Committee, County Offices, Haverfordwest, Dyfed. Telephone: Haverfordwest 3131.

First published 1977

Set in Times 10 on 11 point body and printed by offset lithography.

Cover art work by John Einon.

Design by Brian John.

Printed in Dyfed by Five Arches Press, Knowling Mead, Tenby.

Contents

	Page
Introduction	7
Rock types	8
Igneous rocks	8
Sedimentary rocks	12
Metamorphic rocks	14
Rock Ages	16
Field identification of rocks in the National Park	18
A guide to the identification of folding and faulting	21
Landforms and earth surface processes	26
Erosion surfaces	26
Features from the Ice Age	29
Glossary	32
Some useful references	36

List of Tables

	Page
The geological column in Pembrokeshire	17
Guides to the field identification of rocks	19
(i) Igneous rocks	19
(ii) Sedimentary rocks	19–20
(iii) Metamorphic rocks	20

Frontispiece: Part of the cliff face at Marloes Sands, showing how landforms can be strongly influenced by geological factors. Here the beds of Silurian sandstone and mudstone are standing vertically. The extremely high angles of dip were caused by the exertion of intense pressures during the Armorican orogeny. (Coypright Studio Jon Ltd., Haverfordwest.)

List of Illustrations

	Page
Frontispiece: Vertical beds of rock at Marloes Sands	4
Geological map of Pembrokeshire	6
Maiden's Castle, a crag of rhyolite	9
Columnar jointing in dolerite	10
Pillow lavas at Strumble Head	11
Limestone beds near Stackpole Quay	13
Carboniferous Limestone fossils	14
The rock cycle	15
Diagram of folding in rocks	21
Folded rocks in the cliffs of Dinas Head	22
The Lady Cave anticline, Saundersfoot	22
Fox Hole, a cave at Little Haven	23
Map of the main geological structures of Pembrokeshire	24
Faulted rocks near Broad Haven	25
Erosion surface on Ramsey Island	27
Sands and gravels at Mullock Bridge	30
The Gwaun Valley meltwater channel	31

Acknowledgements

We are grateful to the following for their help in a variety of ways. Photographs have kindly been provided by Studio Jon Ltd., Aerofilms Ltd., and the University of Cambridge Director in Aerial Photography. The maps and dagrams have been drawn by the drawing-office staff of the University of Durham Geography Department and the National Park Department. Gareth Walters and his staff at Five Arches Press have passed the booklet through the press at high speed, and their expertise and generous help has been much appreciated.

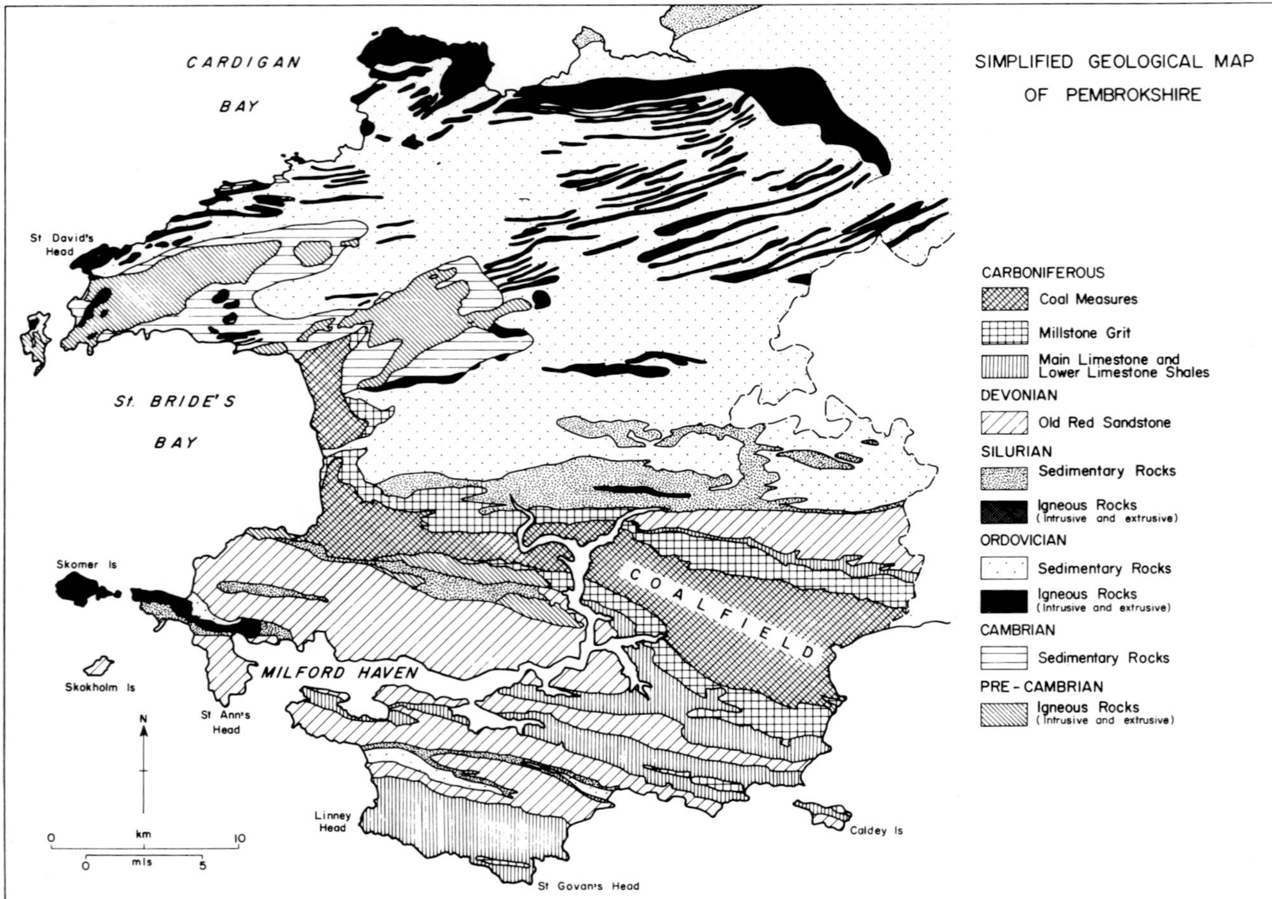

A simplified geological map of Pembrokeshire, showing the arrangement of the main rock types as they

Introduction

The scenery of the Pembrokeshire Coast National Park is made up of four major landform types: coasts,[1] hills, near-flat surfaces, and valleys. These are built of or excavated in rock. To the layman a rock is a solid substance such as a lump of granite, limestone or slate, but to the geologist the loose sands and pebbles of beaches as well as soft clays also qualify as rocks. Any mass of mineral matter which forms part of the earth's crust is a rock, whether it be hard (consolidated) or soft and loose (unconsolidated).

In nature there is an enormous variety of rocks, and one of the most striking features of the geology of the Pembrokeshire Coast National Park is the amazing diversity of types contained in this small area. The rocks range in type from immensely hard crystalline rocks, which have come from deep beneath the earth's surface, to 'superficial deposits' such as blown sands and river gravels. The rocks also range in age from those over 600 million years old to those which have formed within the last decade. There are many excxellent coastal exposures in Pembrokeshire where the enthusiast can examine rocks of all types. Many of these exposures are accessible by car, and many more can now be reached by means of the Long-Distance Coastal Footpath.

This booklet sets out to explain how the main rock types have been formed, with reference to many examples from within the National Park. After a discussion of rock characteristics and rock ages, there are notes on the field identification of rock types and on the recognition of structures such as faults and folds. The final section of the booklet is concerned with landforms, and some guidance is given on the natural processes which have shaped the face of the land particularly over the last two million years. Throughout the booklet important sites are located by means of six-figure grid references, and the reader is advised to refer to O.S. 1:50,000 maps nos. 145 (Cardigan), 157 (St. David's and Haverfordwest) and 158 (Tenby) while looking through the text. Distances are given in both miles and kilometres, and other units of measurement are given with metric equivalents. In case of any difficulty over geological terminology, the reader is advised to consult the glossary at the back of the book, and further enlightenment can no doubt be obtained from looking at some of the items on the short list of references.

1. See booklet No. 2 in this series, entitled *Coastal Scenery of the Pembrokeshire Coast National Park.*

Rock Types

For the purposes of study and identification the rocks making up the earth's crust can be conveniently sub-divided into three main types:

(i) igneous

(ii) sedimentary

(iii) metamorphic

Igneous Rocks

These make up the bulk of the earth's crust and are formed as a result of the hardening or solidification of molten rock material called *magma*. Magma has its source beneath the earth's crust and during phases of volcanic activity it may be forced or intruded into the crust. Alternatively it may actually reach the earth's surface where it is extruded or poured out as molten rock called *lava*. The lava may flow from a fissure or crack in the earth's surface or be emitted from a volcano.

Magma may be intruded into the crust in large or small masses and its solidification leads to the formation of intrusive igneous rock. Solidification of the large masses takes place very slowly and as a consequence large crystals develop. These are the individual minerals which go to make up the rock. Such rocks are coarse-grained, and examples to be found within the Pembrokeshire Coast National Park include: some of the *diorites* exposed in the cliffs to the south-west of Borough Head (SM 846124), the *gabbro* which forms St. David's Head (SM 721278) and Carn Llidi (SM 738280), and the micro-granite (*granophyre*) at Porth Clais (SM 742242). In these rocks the minerals can be identified by the naked eye, the crystals being larger than 1–2mm. in diameter. Solidification of smaller intrusions takes place more quickly and hence the crystals are smaller (0.1 to 2mm.) and are more difficult to recognise with the naked eye. Examples of these medium-grained intrusive igneous rocks include several varieties of *dolerite* which are found for example near Porthgain at SM 807328 and on the summit of Carn Meini (SN 144325).

The solidification of lava results in the formation of an extrusive igneous rock. Under atmospheric pressure lava solidifies rapidly; consequently the resultant rock is characteristically fine-grained, the crystals generally being below the limit of resolution of the naked eye (less than 0.1mm.), although it may be possible to see them with the aid of a good hand lens. Examples of fine-grained extrusive igneous rocks include the *basalt* exposed in the cliffs of Marloes Sands at SM 785074, and at the Coastguard Lookout (SM 755093) near Wooltack Point; and the basaltic *pillow lava* at SM 898414 on Strumble Head. Just outside the National Park boundary

Roch Castle (SM 881212) stands upon a crag of solidified lava called *rhyolite* and this same rock type is again exposed at Maiden's Castle (SM 954248) overlooking Trefgarn Gorge.

The crag of Pre-Cambrian rhyolite at Maiden's Castle, overlooking the northern end of Trefgarn Gorge. (Copyright B. S. John.)

The *pyroclastic* (pyro–fire; clastic–fragmented) rocks are a rather special category of extrusive igneous rock. Volcanic eruptions are often accompanied by violent explosions which result in masses of angular rock fragments, lumps of lava, ash and fine dust being blown into the air. This material is either deposited on the surrounding land surface where it accumulates directly, or on the sea where it gradually sinks and builds up on the sea bottom. In this manner great thicknesses of material may accumulate over very short periods (for example, following the violent eruptions of Vesuvius in Italy in AD 43 the Roman settlements of Pompeii and Herculaneum were entirely engulfed by such detritus). Even areas hundreds of miles away from an eruption may be affected by a 'rain' of very fine ash and dust. When consolidated the coarse debris (rock fragments) gives rise to a hard rock called *agglomerate* whereas the ashes on consolidation give rise to volcanic *tuff*; both types are to be seen at Porth Lysgi (SM 730237). Because of their mode of deposition the pyroclastic rocks possess a number of characteristics similar to those of sedimentary rocks.

In addition to their crystalline nature igneous rocks (excluding the pyroclastic group) possess other physical characteristics which may in turn aid their field identification. Igneous rocks are jointed. Joints are fractures or cracks in the rock and sometimes these display particular, and often striking, patterns. For example, the dolerite forming the headland of Penanglas (SM 948405), to the north of Goodwick, is characterised by *columnar jointing,* the rock mass having been divided into columns or prisms which in cross section are often 4, 5 or 6 sided. These structures develop during the cooling of molten rock; as it cools contraction leads to the production of columns which are perpendicular to the cooling surfaces. (The drying and cracking of mud under the influence of the sun leads to the development of a similar pattern.)

Columnar jointing on the cliffs which form the headland of Penanglas, near Goodwick. (Coypright Studio Jon Ltd., Haverfordwest.)

Pillow lavas at Strumble Head. These lavas were laid down under water during Ordovician times. (Coypright Studio Jon Ltd., Haverfordwest.)

Volcanic lavas sometimes display two particular features; pillow structure and vesicular structure. When lava has been extruded under water as the result of a submarine volcanic eruption, or when it has flowed into the sea from a land-based eruption, rapid cooling takes place. This leads to the formation of distorted or globular masses of lava which when piled one on top of another take on the appearance of a pile of pillows. Examples of basaltic *pillow lava* are exposed at Strumble Head at SM 895414 a short distance east of the derelict Coast Guard Station. During the cooling of lava gases are given off and these produce bubbles within the molten or semi-molten rock. On solidification, many of these gas bubbles become trapped within the rock thereby producing cavities or *vesicles*. Much of the basaltic lava exposed in the cliffs of Marloes Sands at SM 785074 displays this vesicular structure.

Sedimentary Rocks

Although igneous rocks make up the bulk of the earth's crust, sedimentary rocks are the most common types exposed at the surface. Indeed sedimentary rocks, which effectively mask the igneous core, cover over 75% of the earth's land surface. The same pattern is evident within the Pembrokeshire Coast National Park, the bulk of the exposed rocks being sedimentary in type.

The surface of the earth is under continuous attack by agents of erosion such as rain, rivers, sea, wind and moving ice. As a result of action such as the pounding of waves along exposed sea cliffs or the grinding of rocks by glaciers moving down-valley, hard rock is broken down to produce sediment in the form of pebbles, sand, silt and clay. This, in turn, may be transported by rivers, waves, wind or ice, eventually to be deposited layer upon layer in the sea, on the floors of lakes or in river valleys, particularly at river mouths, thereby resulting in the formation of sedimentary deposits. These layered or bedded sediments are transformed into solid rock by a combination of pressure exerted by the weight of overlying sediments and natural *cementation.* In this manner masses of pebbles are formed into a hard rock called *conglomerate,* such as the Cambrian conglomerate exposed at SM 752242 in St. Non's Bay. Sand is transformed into *sandstone,* such as the Ordovician sandstone to be seen at Poppit (SN 147489). And the fine-grained deposits harden into *mudstone* and *shale,* examples of Carboniferous mudstone being visible near Little Haven at SM 858132. Together sandstone and shale account for some 75% of all sedimentary rocks and they are the most commonly occurring rocks within the National Park. In addition to sedimentary rocks composed of fragments of pre-existing rocks there are also organic deposits of which two main types exist: *organic limestone* and *coal.*

Organic limestone is composed chiefly of the skeletal remains of marine organisms such as shells and corals, the sea being the chief environment of deposition. Some limestones may be composed almost entirely of fragmented portions of coral reefs which flourished in clear, shallow seas whereas others are made up largely of whole or fragmented shells. The Carboniferous Limestone which is exposed in the coastal cliffs between St. Govan's Head and Linney Head (SR 975927 and SR 883958) is an excellent example of organic limestone.

Coal is formed chiefly from the remains of plant debris that once flourished as luxuriant, tropical swamp forest. This organic material accumulated and was preserved under waterlogged conditions and subsequently hardened, under pressure, to form rock. Under very great pressure most of the water and gases (resulting from partial decomposition of the plant matter) are expelled and the coal thereby produced is referred to as *anthracite,* a high quality coal very rich in carbon. Anthracite was formerly mined in the Pembrokeshire coalfield at collieries such as the Trefrân Cliff, near Nolton Haven (SM 856197), and thin seams are also exposed in the cliffs between Wiseman's Bridge (SN 146062) and Amroth (SN 163070).

Two outstanding characteristics possessed by sedimentary rocks, and which are related to their mode and environment of deposition, serve to distinguish them from igneous and metamorphic rocks: their bedded and jointed nature, and their contained fossils.

Thin beds of limestone exposed in the cliffs near Stackpole Quay, South Pembrokeshire. (Copyright Studio Jon Ltd., Haverfordwest.)

Layering or *bedding* is of almost universal occurrence in sedimentary rocks but is generally absent in igneous and metamorphic rocks. It is a reflection of the fact that the original inorganic and organic deposits(e.g. sand grains and shells) were laid down more or less horizontally, layer by layer. Each layer, or bed, is separated from the next by a *bedding plane* and the thickness of beds may vary greatly both within and between deposits. Due to rock folding horizontal bedding is rarely encountered and instead the rock layers tilt or dip in various directions. Jointing in sedimentary rocks is developed at right angles to the bedding and where both jointing and bedding are well developed, such as in the Carboniferous Limestone in the south of the National Park, the rock frequently breaks down readily into rectangular blocks.

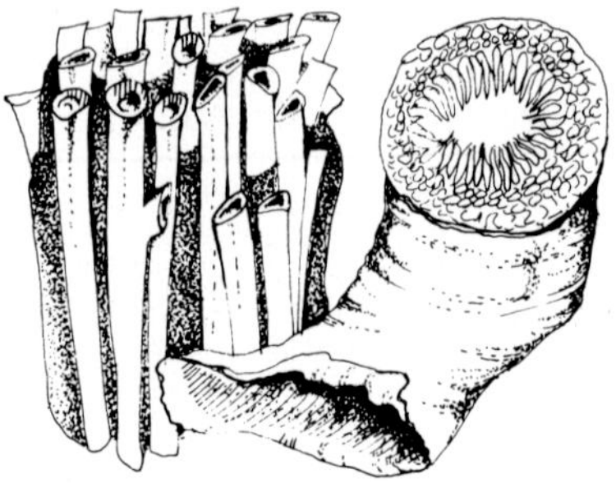

Typical Carboniferous Limestone fossils which can be found in South Pembrokeshire. They are Lithostrotion *(left) and* Caninia *(right). (Copyright National Park Department.)*

Fossils are the whole or fragmented remains of animals or plants that have become preserved within the rock mass. In addition, such indirect evidence of life as animal footprints, burrows and excreta are also classed as fossils. Most sedimentary rocks contain fossils but they are usually absent from igneous and metamorphic rocks. Furthermore fossils are not only diagnostic of sedimentary rocks but the type of fossil, coupled with the evidence provided by rock type, provides an important clue to the environment at the time of deposition and may also indicate the age of the rock.

Metamorphic Rocks

When either igneous or sedimentary rocks are subjected to very intense heat and/or pressure the physical characteristics of the original rock may be altered to such an extent that a new rock type is formed. This *metamorphism* commonly takes place around the margins of igneous intrusions or at great depth within the earth's crust during major periods of earth movement (folding and faulting). In the Pembrokeshire Coast National Park examples of metamorphosed rocks are scarce, the one exception being slate, which results from the alteration of shale or mudstone. Slate was formerly quarried extensively at Rosebush (SN 079303) and at a number of other localities in the Presely Hills.

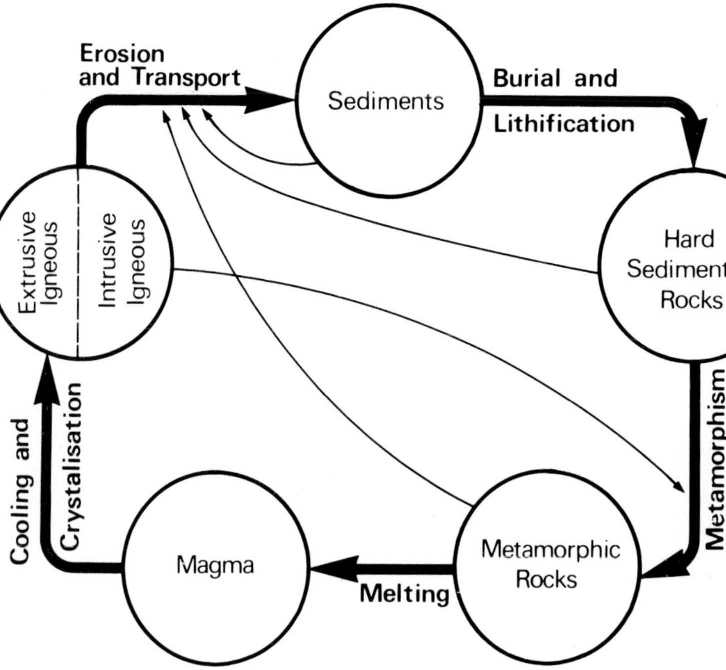

The rock cycle. This shows the relations between the three main types of rock, and the way in which rocks are altered at and beneath the earth's surface. (Copyright National Park Department.)

Although the three major rock types can be treated independently, they are inter-related. For example molten rock material or magma produces either intrusive or extrusive igneous rock; the latter may, in turn, be erodod to form sediments, the raw material of hard sedimentary rock. Deep burial and deformation may result in the transformation of the sedimentary rocks to form metamorphic rocks. The latter, through the application of intense heat, may melt to produce magma. The complete rock cycle is shown by the larger arrows in the accompanying diagram, whereas the smaller arrows indicate possible interruptions or short circuits, e.g. erosion of metamorphic rocks to form sediment.

Rock Ages

Rocks exposed at the earth's surface vary in age from the oldest, which are about 4,500 million years old, to those which are still in the process of formation. This enormous span of geological time has been sub-divided into a number of broad time units called *eras* and narrower units called *periods*; the latter are further sub-divided into *epochs*. Furthermore, on the basis of the evidence provided by the relationship of rock masses one to another, their contained fossils, and direct or radiometric dating any rock (be it sedimentary, igneous or metamorphic) can be assigned to a particular geological period or chapter in earth history. The accompanying table shows how the rocks of Pembrokeshire are arranged on the "geological column." Within the area encompassed by the National Park, and indeed within south-west Wales in general, the geological record is very incomplete. The oldest, or Pre-Cambrian, rocks are probably no older than 2,000 million years whereas the newest consolidated rocks are 250 million years old. There are no solid rocks younger than the Coal Measures, although there are some mysterious "pipeclays" near Flimston (SM 927954) which may be only about 26 million years old. The unconsolidated sands, gravels, clays and other materials which cover the solid rock over wide areas are of course geologically very recent. Most of them are less than 100,000 years old.

THE GEOLOGICAL COLUMN IN PEMBROKESHIRE, SHOWING ROCK AGES AND OTHER FEATURES OF INTEREST

Era	*Age in millions of years*	*Period*	*Epoch*	*Main rock types and environment of deposition*	*Some characteristic fossils*
CENOZOIC		Quaternary	Pleistocene	unconsolidated sands, gravels and clays of glacial, marine, and fluvial origin.	land plants and sea shells
	2	Tertiary	Pliocene	wholly unrepresented ALPINE OROGENY	
			Miocene		
	26		Oligocene	almost wholly unrepresented	
			Eocene		
			Palaeocene		
MESOZOIC	65	Cretaceous		almost wholly unrepresented	
		Jurassic			
		Triassic			
PALAEOZOIC (UPPER)	225	Permian		wholly unrepresented	
	280			ARMORICAN OROGENY	
		Carboniferous	Coal Measures (Westphalian)	fluvial sandstone and shale and coal	land plants and freshwater shells
			Millstone Grit (Namurian)	mainly fluvial sandstone and shale	land plants; ammonites concentrated in thin marine bands
			Carb. Limestone (Avonian)	marine organic limestone	abundant sea shells, corals and sea lilies
	345	Old Red Sandstone (O.R.S.)	Upper Middle Lower	mainly fluvial (river) sandstones, marls and conglomerates. CALEDONIAN OROGENY	
PALAEOZOIC (LOWER)	395	Silurian	Ludlow Wenlock Llandovery	mainly marine shales and gritty sandstones; also much intrusive and extrusive igneous rock	corals and sea shells
	440	Ordovician	Bala Llandeilo Llanvirn Arenig	mainly marine shales and gritty sandstones; also much intrusive and extrusive igneous rock	graptolites and trilobites
	500	Cambrian	Upper Middle Lower	mainly marine sandstones with shales	trilobites and some sea shells
EOZOIC	570	Pre-Cambrian	Pebidian Johnston	intrusive and extrusive igneous rocks	

1, 2 These are "time labels". The equivalent "stratigraphic labels" are *system* and *formation*.

Field Identification of Rocks in the National Park

The detailed identification of rocks is a highly specialised pursuit and much of the work has to be carried out in the laboratory. This is particularly so in the case of igneous rocks, the recognition and naming of which involve assessments of crystal size and type and of the relative amounts of constituent minerals. Such work involves the cutting of thin sections of rock and their painstaking analysis under a geological microscope. Nevertheless by careful observation, and on the basis of field evidence alone, the layman should find it possible to identify the major rock types encountered within the Pembrokeshire Coast National Park. In most cases consultation of the geology map of the area will then help you to tell the age of a rock. No tools more elaborate than a hand lens, a pocket knife and small geological hammer are necessary. A hammer may be required in order to obtain a fresh, clean and unweathered rock face for closer examination. However, on no account should *in situ* rock faces be the subject of hammer blows; many of the rock exposures within the Park, as elsewhere, are of great scientific value and require protection. Be content, therefore, with the examination of broken rock fragments that usually litter the base of exposures (your attention is also drawn to the extracts from the *Code of Geological Conduct* which is printed inside the back cover of this booklet). An additional item which can greatly aid rock identification (particularly limestones) is a small bottle of dilute hydrochloric acid, but use it sparingly and with care, ensuring that it is kept well out of the reach of children.

A GUIDE TO THE FIELD IDENTIFICATION OF THE MAJOR ROCK TYPES OF THE PEMBROKESHIRE COAST NATIONAL PARK (N.B. this should not be used uncritically elsewhere.)

(i) IGNEOUS ROCKS

General characteristics: (i) crystalline texture (coarse, medium or fine). (ii) because of their mineral content most of the dark-coloured igneous rocks of the area are heavier in weight than comparable sized pieces of sedimentary rock. (iii) tend to be very hard and difficult to break even with the aid of a geological hammer. (iv) generally unbedded, but jointed, and unfossiliferous.

Rock Type	*General Description*	*Colour*	*Examples of particular varieties with locations*	*Probable environment of deposition/ solidification*
micro-granite (granophyre)	medium to coarsely crystalline: dominantly composed of minerals belonging to the felspar group.	grey-white to pinkish grey-white with black specks.	granophyre (Pre-Cambrian): Porth Clais (SM 741243) and headland at SM 731234 east of Porth Lysgi.	large scale instrusions of magma: cooled slowly deep in earth's crust.
gabbro/ diorite	medium to coarsely crystalline gabbro is composed dominantly of plagioclase felspar (white mineral) and pyroxene (dark mineral) whereas diorite is of plagioclase felspar and hornblende (dark coloured mineral).	very dark coloured greenish-grey or greenish black and speckled white. Diorite tends to be blacker.	diorite (Pre-Cambrian): Borough Head (SM 846124) gabbro: (Ordovician) St. David's Head (SM 721278)	Large scale intrusions: magma cooled slowly deep in earth's crust.
dolerite	medium crystalline; mineral content same as that for gabbro but hand lens required to make out crystal boundaries. Sometimes displays a porphyritic texture, that is where large crystals (or phenocrysts) are set in a finer crystalline ground mass: the spotted dolerite of Carn Meini is a good example. Also occasionally displays columnar jointing.	dark bluish-grey green	dolerite (Ordovican): Penanglas (SM 949405)–displays columnar jointing dolerite (Ordovician): Carn Meini (SN 144325); phenocrysts of felspar.	medium to small size intrusions cooling more rapidly than in the case of gabbro.
basalt	finely crystalline; generally no minerals identifiable in hand specimen even with aid of hand lens. Mineral content similar to dolerite and gabbro. Sometimes displays pillow and/or vesicular structure.	black, greyish-black or grey-green.	basalt (Ordovician): Strumble Head (SM 898414); displays pillow and vesicular structure. basalt (Silurian): Marloes Sands (SM 785074); displays vesicular structure.	rapid cooling of lava under atmospheric conditions
rhyolite	fine to very crystalline with no minerals identifiable in hand specimen even with aid of hand lens. Mineral content mainly felspar and quartz with some mica.	light coloured: most commonly white, grey or reddish-brown.	rhyolite (Pre-Cambrian): reddish-brown; Maiden's Castle (SM 954248) rhyolite (Ordovician): grey; Pwll Lleuog (SM 731275)	rapid cooling of lava of high viscosity under atmospheric conditions.
tuff and agglomerate	Tuff is composed of volcanic fragments generally less than 2mm in diameter; some larger, usually angular, fragments may be present. Texture may be similar to a sandstone. Agglomerates are composed of a mixture of angular to subrounded fragments set in a finer ground mass.	variable; including buff, yellow, purple, green-grey, pink.	tuffs and agglomerates (Pre-Cambrian): multi-coloured (yellow and purple prominent): Porth Lysgi (SM 730237) rhyolitic tuff (Ordovician): grey; Pwll Lleuog (SM 731275) rhyolitic agglomerate (Ordovician): grey-white; Porth Maen Melyn (SM 888394)	fall-out on land or water of debris blown out of volcanic crater during violent eruption.

(ii) SEDIMENTARY ROCKS

General characteristics: (i) Generally well-bedded (thin or massive) and jointed. (ii) fossiliferous, especially limestones, but bulk of Carboniferous sandstones and O.R.S. sandstone, conglomerate and marl are unfossiliferous or only poorly fossiliferous.

Conglomerate	consists of pebbles upwards of 2mm in size and displaying varying degrees of roundness set in a dominantly sandy ground mass or matrix.	red to reddish brown	Red Conglomerate (Cambrian)–contains white and red pebbles: St. Non's Bay cliff top at SM 754243	shallow sea water (beach deposit)
			Ridgeway Conglomerate (O.R.S.): reddish brown in colour: near West Angle at SM 853038	river bed/delta (gravels)
			Occasional conglomerate layers (Silurian): Marloes Sands (SM 788073)	shallow sea water (beach deposit)

(ii) SEDIMENTARY ROCKS (continued)

Sandstone	consists of sand grains cemented together; deposits usually well-sorted, that is the grains in a particular sandstone are all about the same size; texture varies from fine sandstone (grains) about $\frac{1}{16}$ mm diam.) to coarse sandstone (2mm in diam.) The coarser varieties have a 'gritty' texture. Sometimes bedding planes are ripplemarked in the same fashion as modern beaches, e.g. Silurian sandstones at Marloes Sands.	very variable: may be purple; green; red-brown; yellow-brown; grey-brown or blue-grey.	Purple (Caerbwdy) sandstone (Cambrian): Porth Clais (SM 744238)	shallow sea water
			Green sandstone (Cambrian): Caerfai Bay (SM 760244)	shallow sea water
			Blue-grey gritty sandstone (Ordovician): Poppit (SN 147489)	deep sea water
			grey, brown and yellow-green sandstone (Silurian): Marloes Sands (SM 786074)	shallow sea water
			red-brown sandstone (O.R.S.): Manorbier (SS 059976) and St. Bride's Haven (SM 802110)	river delta/flood plain
			grey-brown sandstone (Carboniferous): Little Haven (SM 856131)	river delta
Shale or Mudstone	very fine grained deposits; grains less than $\frac{1}{16}$ mm diameter and not identifiable except for the sandy-mudstones or sandy-shales even with the aid of a hand lens. If the deposit is very finely bedded so that it splits readily with the aid of a pen-knife into thin layers it is called a shale; when consolidated and more massive (i.e. resists splitting into thin layers) it is called a mudstone. A calcareous mudstone is called a marl.	mostly grey-black, grey or blue-grey. Also grey brown and red.	black shales and mudstones (Cambrian): Porth-y-rhaw (SM 786243); includes fossils (trilobites).	deep sea water
			red shales (Cambrian); called shale but often more reminiscent of mudstone: Caerfai Bay (SM 762243)	deep sea water
			black shales (Ordovician): Abereiddi (SM 796312); fossil graptolites in beach pebbles.	deep sea water
			blue-grey mudstone (Silurian): Marloes Sands (SM 787072); layers of fossils, especially corals.	relatively deep sea water
			red marls (O.R.S.): Freshwater West–Little Furzenip Headland (SR 885994)	river flood plain.
			black mudstone (Carboniferous): Settlands Bay, Little Haven (SM 858132); abundant plant remains.	river delta/lagoonal swamp
Organic Limestone	rock composed of fragmented and whole remains of marine organisms. Rock commonly visibly packed with fossils. The latter, and hence the rock, are composed largely of Calcium carbonate ($CaCO_3$) which will fizz (effervesce) immediately on application of dilute hydrochloric acid. The mineral Calcium carbonate is also soft and hence the rock is easily scratched with a pen knife. Often displays a crystalline texture.	grey blue-grey grey-white	Carboniferous Limestone: continuously exposed in cliffs between Stackpole Quay (SR 993958) westwards to Linney Head (SR 883958); also between Skrinkle Haven (SS 080973) and Tenby (SN 138005); also West Angle (SM 851035). Highly fossiliferous, especially sea shells, corals and sea lilies.	relatively warm, shallow clear sea.
Coal	Not well exposed but thin seams are to be seen interbedded with Carboniferous (Coal Measures) sandstone in the cliffs between Wiseman's Bridge (SN 146062) and Amroth (SN 162070)			tropical lagoonal swamp

(iii) METAMORPHIC ROCKS

slate	similar to shale except that it is generally harder and splits very readily along well-defined preferred directions - called slaty cleavage.	blue-grey to grey-black	slate (Ordovician): Rosebush (SN 089303)	product of metamorphism of shale during periods of earth movements.
			slate (Ordovican): Abereiddi (SM 795315)	

A Guide to the Identification of Folding and Faulting

The hard rocks of the Pembrokeshire Coast National Park are of ancient origin, and even the youngest (the Coal Measures of Carboniferous age) are over 280 million years old. Although these rocks are magnificently exposed in the coastal cliffs, unravelling the 'record of the rocks' is made difficult due to the fact that they are severely folded and faulted. Folds and faults are the products of crustal movements called *orogenies*, or periods of mountain building. Since the Pre-Cambrian there have been three main periods of mountain building, two of which strongly affected the rocks of the National Park. The Caledonian Orogeny of about 400 million years ago buckled and fractured rocks of Cambrian, Ordovician and Silurian age, while the Armorican Orogeny of about 290 million years ago folded and faulted the newer rocks of O.R.S. and Carboniferous age.

As with rock types, the detailed study of folds and faults is highly specialised. Nevertheless, by simple observation, the layman can easily identify the presence of folding and faulting and offer simple interpretations for some of the features observed. The layering (bedding) of undisturbed sedimentary rocks is more or less horizontal. However, the rocks of the Pembrokeshire Coast National Park are everywhere folded and the folds are of two basic types: upfolds (*anticlines*) and downfolds (*synclines*). Folding always results in the tilting of strata. If the pressures exerted are gentle the tilt or dip of the rock layers comprising the flanks or limbs of the folds will be slight. Strong folding on the other hand, which is characteristic of Pembrokeshire, it characterised by high angles of dip.

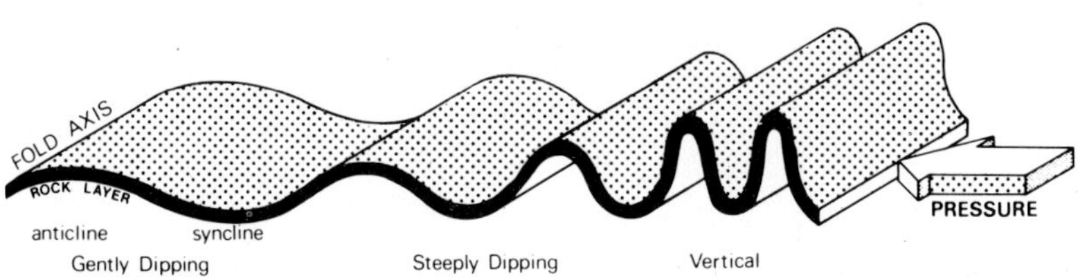

Folding in rocks. The steepness of the dip varies according to the amount of pressure exerted and the distance from the pressure source. (Copyright National Park Department.)

Folded rocks in the Pembrokeshire cliffs. Above, spectacular folds in the cliffs of Dinas Head. The rocks are Silurian mudstones and sandstones. Below, the Lady Cave anticline, Saundersfoot. This well-known feature has been formed in Coal Measures sandstones and shale. (Both photos copyright Studio Jon Ltd., Haverfordwest.)

Sometimes folds are small enough to be seen in their entirety and along the foreshore at Poppit, between SN 144490 and SN 149489, a succession of simple synclines and anticlines in the shales and grits of Ordovician age can be seen in three dimensions. In the nearby cliffs between Pen-yr-Afr (SN 118487) and Cemais Head (SN 132502) the folding is on a larger scale, and only the outlines of the anticlines and synclines are evident in the high cliff faces. North of Broad Haven beach the anticlinal structure known as the Sleek Stone (SM 860144) can be studied in three dimensions whereas further south the Carboniferous sandstone composing the bulk of the headland (SM 856131) separating Little Haven from The Settlands is thrown up into a large anticline. Some of the softer mudstone in the core of this anticline has been removed by wave erosion

The Fox Hole, Little Haven: a cave formed in the core of an anticline in Coal Measures. (Copyright National Park Department.)

to form the cave known as Fox Hole. In a small cove (SM 858135) between The Settlands and Broad Haven thinly bedded Carboniferous sandstones and shales have been very intensely folded, giving rise to a series of small scale anticlines and synclines. To the south of Milford Haven the folding displayed at West Angle Bay (SM 853033) is on a larger scale. Layers of Carboniferous Limestone on the north side of the bay are tilted southwards whereas those on the south side of the bay dip northwards; to stand in the middle of the bay is therefore to stand in the core of a downfold or syncline, the rocks rising upwards to both north and south. A similar pattern is evident at Lydstep Haven. On the north side of the bay the Carboniferous Limestone around Proud Giltar (SS 102985) dips steeply towards the south whereas on the south side the limestone beds composing Lydstep Point (SS 095975), and which represent the southern limb of the downfold, are arranged vertically. The opposite arrangement of rock layers is evident in the neighbouring bay of Freshwater East (SS 019978). Here, the O.R.S. rocks on the north side of the bay dip steeply northwards whilst those on the south side dip less steeply southwards. To stand in the centre of the bay is therefore to stand within the core of a large, unroofed anticline.

In some instances, however, the scale of folding is so large that only one limb of an upfold or downfold is evident. For the most part the Cambrian sandstones and shales exposed in the cliffs between Porth Clais (SM 742238) and Porth-y-rhaw (SM 786243) dip steeply seawards and these rocks represent the southern limb of a massive anticline which overtops the whole of the St. David's peninsula. The northerly dipping rocks forming the northern limb of this upfold are visible in the cliffs at the northern end of Whitesand Bay (SM 733273).

The line which follows the crest of an anticline or the trough of a syncline is called the *fold axis*. This axis is aligned at right angles to the direction from which pressure was applied during the formation of the fold. Hence, in the case of relatively small folds, where it is possible to locate fold axes in the field such as at Poppit, the Sleek Stone (Broad Haven), Fox Hole (Little Haven), West Angle Bay, Freshwater East, and Lydstep it is possible to deduce pressure directions. Indeed, as a result of such observations geologists have found that the folds affecting Cambrian, Ordovician and Silurian strata of the area, which were produced by the Caledonian Orogeny, have a general trend from ENE to WSW whereas later folds brought about by the Amorican Orogeny (and affecting the O.R.S. and Carboniferous rocks) trend ESE to WNW.

THE STRUCTURE OF PEMBROKESHIRE

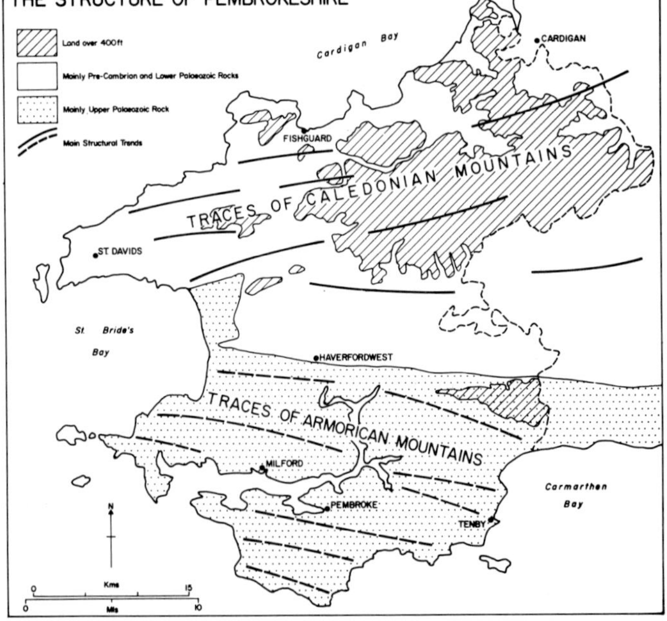

A reverse fault exposed in the cliffs near Broad Haven. The cliff is about 15m. high. (Copyright National Park Department.)

Rock folding experiments have shown that by increasing the pressure, a point is eventually reached when the folded materials cannot endure the strain and consequently fracture. Rock fractures are known as *faults* and the rocks on either side of the line of fracture may be displaced or offset by a matter of millimetres or several kilometres. There are two basic types of fault, normal and reverse (or thrust), although it is often very difficult for the layman to separate them in the field. Fine examples of small scale normal faulting are found in the cores of some of the anticlines along the foreshore at Poppit, whereas to the north of Broad Haven at SN 860144 there is a magnificent example of a reverse (thrust) fault disturbing the folded strata.

Opposite: the two main structural "provinces" of Pembrokeshire. Caledonian structures are found for the most part in the north, and Armorican structures in the south. (Copyright B. S. John.)

Landforms and Earth Surface Processes

Erosion surfaces

With the exception of the Presely Hills and other isolated hill masses the landscape of the Pembrokeshire Coast National Park and adjoining areas is dominated by extensive, near-flat surfaces. The broad, flat-topped coastal zone is frequently referred to as the '200 foot platform.' This term, however, is misleading and it is more precise, though still a generalisation, to speak of *three* near-flat surfaces separated one from another by steps or areas of steeper slope. The lowest surface is at an altitude of about 200 feet but commonly varies between 100 and 250 feet (46–76m.); the '400 foot surface' similarly varies between 350 and 450 feet (107–137m.); whilst the highest '600 foot surface' varies between 550 and 650 feet (168–198m.).

The '200 foot platform' is the most extensive of the three surfaces. South of Milford Haven it is magnificently developed, at an altitude of 100–150 feet (30–46m.), between Stackpole Head (SR 995942) and Linney Head (SR 884957). It cuts quite indiscriminately across highly folded limestone and shale of Carboniferous age. Between Freshwater East and Lydstep Haven parts of the plaftorm, cut across folded O.R.S. and Carboniferous rocks, exceed 200 feet (61m.) O.D., and offshore the pattern is continued on the remarkably flat-topped island of Caldey. North of the Haven the surface, planed across folded and faulted O.R.S. and Silurian sandstones and mudstones and extrusive igneous rocks of Silurian age forming the Dale-Marloes peninsula, varies in elevation between 150 and 250 feet (46–76m.), having an average height of about 200 feet (61m.). The platform developed on the St. David's peninsula and offshore Ramsey is at a similar elevation and its monotony is broken only by upstanding craggy hill masses such as Carn Ysgubor (SM 700247), Carn Llundain (SM 696234), Carn Llidi (SM 738280), Carn Ffald (SM 751283) and Penbiri (SM 767293) and smaller but no less dramatic rock outcrops such as Carn Treglemais (SM 816281). On the St. David's peninsula this surface cuts across Cambrian sedimentary and Pre-Cambrian igneous rocks. To the north the 200 foot surface is again evident, planed across the extrusive igneous rocks of Ordovician age which underlie the Pen Caer peninsula.

The 400 foot surface is less extensive than the 200 foot platform. However, it is quite well developed between Croesgoch (SM 828303) and Mathry (SM 879320), where it cuts across folded and faulted shales and grits and extrusive rocks of Ordovician age. South-east of Croesgoch

The coastal platform magnificently developed on Ramsey Island, off the tip of St. David's Peninsula. The former islands (monadnocks) can be clearly seen standing above the erosion surface. (Copyright Aerofilms Ltd.)

an extensive remnant of the 400 foot surface is found around Hayscastle Cross (SM 913252) and it is overlooked from the south by Dudwell Mountain (SM 907232) and Great Trefgarn Mountain (SM 945243). The same surface is found south and east of Fishguard, forming the summit of Dinas Island (SN 005407).

Within the National Park the 600 foot surface is only found to the north-east of Newport. In this area the plateau cuts across intensely folded and faulted shales and grits of Ordovician age and lies, for the most part, between 550 and 600 feet (168–183m.). Beyond the Park boundary it is well developed as a broad fringe around the southern margin of the Presely Hills, particularly in the vicinity of Maenclochog (SN 083274), Llangolman (SN 118269) and Efailwen (SN 134253). It is also interesting to note that hill summits such as Carn Llundain, Carn Llidi, Garn Fawr (SM 896388), Garn Gilfach (SM 909391), Carn Gelli (SM 923379), Dudwell Mountain and Great Trefgarn Mountain rise to about 550–650 feet and hence may be dissected remnants of the same surface.

The near-flatness of the 200, 400 and 600 feet surfaces contrasts strongly with the complexity of the bedrock. Irrespective of differences in rock type, rock age and folding and faulting the general plateau form is maintained over large areas and we are left with the conclusion that the regional landscape of the Pembrokeshire Coast National Park is only in minor degree a reflection of rock type and structure.

The one agent of erosion capable of planing a near-flat surface indifferently across complex geological structures is the sea. As the sea advances across a land surface the coastal cliffs gradually retreat under the influence of wave attack and at their base is left a *wave-cut rock platform* which, around present-day shorelines, is often masked by beach deposits: pebbles, shingle and sand. Cliffs and associated wave-cut platforms are common phenomena around the coastline and particularly extensive platforms fringe the cliffs between Amroth (SN 162069) and Saundersfoot (SN 138048) and at Freshwater West, between Little Furzenip (SR 884994) and Great Furzenip (SR 887987) headlands. The 200, 400 and 500 foot surfaces are interpreted as the remnants of ancient wave-cut rock platforms cut when sea-level stood higher than it does today, whilst the steps between the platforms are regarded as ancient cliff-lines. In places, the continuity of a surface is broken by isolated hill masses and these are interpreted as *monadnocks* or former islands. For example Carn Ysgubor, Carn Llundain, Carn Llidi, Carn Ffald and Penbiri overlook the 200 foot surface cut across the St. David's peninsula, and Garn Fawr and Garn Gilfach rise above the same surfaces developed across Pen Caer. All of these formed islands when the sea level stood at about 200 feet. Similarly Dudwell Mountain and Great Trefgarn Mountain formed islands when the sea level stood at 400 feet, in the same way that Ramsey and other offshore islands stand above the present sea level. These ancient islands, like present-day examples, are generally composed of hard, resistant rock which yields less readily under wave attack. Carn Llidi, Carn Ffald, Penbiri, Garn Fawr and Garn Gilfach, for example, are monadnocks composed of resistant intrusive igneous rocks of Ordovician age. Dudwell Mountain and Great Trefgarn Mountain are built of hard Pre-Cambrian extrusive igneous rocks. More subtly the 200 foot platform which bevels the Dale-Marloes peninsula rises imperceptibly to almost 300 feet south of Borough Head (SM 837125) and it is significant to note that this area of higher ground is underlain by harder and more resistant intrusive rock of Pre-Cambrian age.

The surfaces are apparently devoid of any covering of beach material, which has presumably been removed by later erosion. This makes dating extremely difficult, although there is evidence to suggest that the 600 foot surface, the highest and oldest, was being fashioned by the sea no more than 2 million years ago. When the sea level stood at 600 feet the shoreline lay close to the base of the Presely Hills. At a later date the sea level fell to about 400 feet, and the 400 foot surface developed at the expense of the 600 foot. Similarly, the 200 foot surface developed at the expense of the 400 foot. As a consequence of the gradual fall in sea level rivers extended across the newly emerged land areas, and these were responsible for cutting many of the valleys which today break the continuity of the ancient wave-cut surfaces.

Features of the Ice Age

Some of the valleys which dissect the hill and plateau areas were cut under very special circumstances. These are *glacial meltwater channels*, so called because they were excavated by meltwater escaping from melting ice sheets which once overwhelmed the whole of Wales. These channels, together with a number of other landscape features within the National Park, have probably been formed within the last 250,000 years—that is, very recently in geological terms.

The Pleistocene epoch is often referred to as the Ice Age and there is evidence to suggest that south-west Wales suffered glaciation on at least two occasions, between which times the climate was more like that experienced today. Such an intervening warm phase between glaciations is referred to as an *interglacial*. Evidence within the National Park for the earlier glaciation, which probably occurred more than 100,000 years ago, is scanty. Nevertheless foreign boulders or *erratics*, which were transported great distances southwards by the glaciers generated in the highland areas of the British Isles, are found scattered over the territory to the south of Milford Haven. In contrast to the earlier glaciation, when ice overwhelmed the entire area, the Last Glaciation probably affected only the areas to the north of the Haven. About 18,000 years ago St. Bride's Bay, the St. David's peninsula, the area south of Fishguard and the territory north of the Presely Hills was invaded by a great glacier which had moved southwards down the Irish Sea basin. This glacier is referred to by geomorphologists (landform scientists) as the Irish Sea Glacier.

In constricted valleys in highland areas such as Snowdonia, ice may effect a great deal of erosion. But in lowland Pembrokeshire little erosion was achieved, except for some scratching *striation* of rock surfaces such as is found on the flanks of Carn Llidi at an altitude of about 500 feet (152m.). In contrast, however, the ice deposited much *till* or boulder clay, and this material, with its contained erratics which in some cases are derived from as far away as Scotland, is widely spread over the plateau surfaces. Also, it partially fills some of the valleys such as at Abermawr (SM 882345) and Druidston Haven (SM 862172). The brownish-purple/blue till, like the associated deposits of sands and gravels, contains sea-shell fragments dredged off the floor of the Irish Sea as the ice moved southwards. Indeed, in places the till is highly calcareous, being composed largely of old sea-floor materials. As a consequence samples of till will fizz (effervesce) on application of dilute hydrochloric acid.

As the climate became warmer when the Last Glaciation was drawing to an end, the ice began to melt and the meltwater so produced was in places responsible for the deposition of sands and gravels. These unconsolidated deposits are often found as rather featureless spreads in association with till, as at Abermawr and Druidston. Alternatively they

Sands and gravels laid down by meltwater at the end of the last glaciation at Mullock Bridge, near Dale. The photo shows some excellent examples of faulting (Copyright B. S. John.)

may form spectacular upstanding mounds, ridges and terraces. Outside the National Park boundary the quarried mounds of sand and gravel at Banc-y-Warren (SN 204476), near Cardigan, are very conspicuous. At Monington (SN 139438) the sands and gravels, deposited by a river flowing beneath the ice, form a sinuous ridge or *esker* whereas at Mullock Bridge (SM 811078), near Dale, they are banked up in the form of a terrace and were probably deposited in standing water which had accumulated on the side of an ice mass occupying the Gann valley.

However, the most spectacular landforms of glacial origin in this area are those created by the erosive action of meltwater. The most impressive meltwater channel is the Gwaun Valley which separates Carn Ingli (SN 062372) from the main mass of the Presely Hills. This, together with the set of channels south of Fishguard, forms a complex interlinked system quite unlike normal river valleys, and their scale of development is out of all proportion to the small streams which now occupy them. Sometimes streams may be absent from these valleys, as in the case of the valley separating Dinas Island from the mainland. Other examples of meltwater channels include: that linking Porthgain (SM 814326) with Abereiddi (SM 797313) and isolating Ynys Barri; Merry Vale (SM 745250); the Solfach valley (SM 806244); and the St. Elvis Valley (SM 810243). All these are interpreted as having been cut by huge volumes of meltwater flowing beneath melting ice masses. Some may date from the earlier glaciation although many were doubtless cut or greatly modified during the melting phase of the Last Glaciation.

By 15,000 years ago climatic amelioration had almost completely freed the area of ice although the last permanent ice masses did not finally leave the mountain areas of the British Isles until 10,000 years ago. During glacial periods enormous quantities of water are locked up in ice bodies and as a consequence sea-level falls. On melting, this water is released into the ocean basins and as a result sea level rises again; after the Last Glaciation sea-level did not reach its present position until about 6,000 years ago. This rise in sea-level led to many of the most recent landscape changes which have taken place in this area. For example, we can cite the drowning of ancient forests, and the flooding of deep river valleys such as Milford Haven. Since these features are found around our coastline they are more fully described in *Coastal Scenery of the Pembrokeshire Coast National Park*.

The Gwaun Valley, one of the most impressive meltwater channels in the British Isles. Notice how the channel is incised deeply into the old land surface. (Cambridge University Collection; copyright reserved.)

The fascinating paradox presented by a study of the rocks and landforms in and around the Pembrokeshire Coast National Park is that while the hard rocks are very ancient—none are younger than 250 million years—the landscape is a compound of landforms, chiefly the product of erosion by sea and rivers (including glacial meltwater), which are of very recent origin, probably for the most part no more than 2 million years old.

Glossary

AGGLOMERATE: A mass of coarse rock fragments or angular blocks of lava produced during a volcanic eruption.

ALPINE OROGENY: The great mountain-building phase which occurred during Miocene times. Responsible for the building of the Alps, Rockies, Andes, etc.

ANTICLINE: An upfolded or arched structure in bedded rocks, generally produced as a result of pressure from the sides.

ARMORICAN OROGENY: The mountain-building episode at the end of Carboniferous and in early Permian times. (From Armorica, that part of France now called Brittany.)

ASH: The fine material thrown out during a volcanic eruption.

BASALT: A dark, fine-grained lava. The commonest of all lava types.

BEDDING PLANE: The surface which separates one layer of sedimentary rock from another.

CALEDONIAN OROGENY: The great mountain-building episode of late Silurian and early Devonian times. (From Caledonia, or Scotland, where most of the mountains are of this age.)

CAMBRIAN: The earliest geological period of the Palaeozoic; lasted from 570 to 500 million years ago. The first geological system to yield abundant fossils. (From Cambria, an old name for Wales.)

CARBONIFEROUS: Literally "coal-bearing." This period lasted from 345 to 280 million years ago, and is classified as part of the Upper Palaeozoic era.

CEMENTATION: The hardening of a rock as a result of chemical changes in the matrix.

CENOZOIC ERA: The "recent life" era, lasting from about 65 million years ago to the present. It includes the Tertiary and Quaternary periods.

COAL: Rock formed from peat or other vegetable matter after a long period of compaction.

COLUMNAR JOINTING: Jointing in an igneous rock resulting from cooling, and giving rise to "columns" perpendicular to the cooling surface.

CONGLOMERATE: A rock made up of a mass of pebbles or rounded boulders set in a matrix of finer material.

CORAL: A hard stone of calcium carbonate, made up of the skeletons of millions of small marine animals.

CRETACEOUS: The final period of the Mesozoic era, ending about 65 million years ago.

CRYSTALLINE ROCK: Rock made up of varied mineral crystals which have formed as a result of cooling. The term is usually used for igneous rocks.

DEVONIAN: The geological period which followed the Caledonian Orogeny and lasted for about 50 million years. (Named after Devon, where rocks of this age are well exposed.) Also called "Old Red Sandstone."

DIORITE: A coarse-grained intrusive igneous rock. Often dark in colour.

DOLERITE: A dark-coloured igneous rock, resembling a basalt but composed of larger crystals.

EOCENE: The second epoch of the Cenozoic era. Lasted 15 million years.

EOZOIC: The earliest geological era, lasting from the creation of the earth (about 5000 million years ago) until about 570 million years ago.

EPOCH: The term used for a subdivision of any of the geological periods. Hence the Coal Measures were laid down during the final epoch of the Carboniferous period.

ERA: The longest unit of geological time. The Palaeozoic era, for example, lasted for 345 million years.

EROSION SURFACE: An undulating surface cut across variable bedrock by natural agencies such as running water, waves or glaciers.

ERRATIC: A boulder or rock fragment transported some distance from its original bed, generally by the action of glacier ice.

EXTRUSIVE ROCK: Rock resulting from the solidification of lava or other volcanic materials at the ground surface.

FAULT: A dislocation or fracture in the rocks, where one side has moved relative to the other.

FOLD: A general term for any flexure or bend in the rocks caused by compression in the earth's crust.

FORMATION: Group of rocks which can be recognized as belonging to a certain geological epoch.

FOSSIL: Anything found in a rock which is recognizable as the remains of a plant or animal.

GABBRO: An intrusive igneous rock, usually coarse and crystalline.

GLACIAL STAGE: Any part of the Pleistocene during which glaciers cover large parts of the continents. The next one is due in about 10,000 years' time.

GRANITE: A coarse-grained igneous intrusive rock, often white or pink in colour.

GRIT: A coarse type of sandstone, composed of large and sometimes angular grains of quartz.

GRANOPHYRE: A light coloured crystalline igneous rock. Crystals are smaller than in granite, and for this reason it is also known as "microgranite."

HOLOCENE: The most recent division of geological time, lasting from 10,000 years ago to the present. A part of the Quaternary period.

IGNEOUS ROCK: A rock formed by the solidification of molten material from the earth's interior.

INTERGLACIAL STAGE: Any part of the Pleistocene epoch during which there is relatively little glacier ice present on the earth's surface. The Holocene of the present day is one such interglacial.

INTRUSIVE ROCK: A rock which has been forced, while in a molten state, into the cavities or cracks in pre-existing rock strata.

JOINT: A line of breakage or fracture which cuts across a rock structure and allows its separation into blocks.

JURASSIC: The middle period of the Mesozoic era. The time of dinosaurs.

LAVA: Molten igneous rock which has flowed from the earth's interior (often during a volcanic eruption) to cool and solidify at the surface.

LIMESTONE: Rock rich in calcium carbonate. Usually formed in clear, warm water from the remains of marine organisms such as coral and sea shells.

MAGMA: The molten material from the interior of the earth from which all igneous rocks are made.

MESOZOIC: The "middle life" era, lasting from 225 to 65 million years ago.

METAMORPHIC ROCK: Rock which has been greatly changed by heat or pressure.

MINERAL: A solid inorganic substance which has recognizable physical and chemical properties. Rocks are usually mixtures of minerals.

MIOCENE: One of the epochs of the Tertiary period. A time of intense earth movements—usually referred to as the Alpine Orogeny.

MONADNOCK: An upstanding rocky hill, often an ancient island standing above a wave-cut platform.

MUDSTONE: A fine-grained sedimentary rock, usually formed of silt and clay laid down in deep water.

OLD RED SANDSTONE: The geological period which followed the Caledonian Orogeny, about 395 million years ago. As the name implies. many of the O.R.S. rocks are sandy and red-stained.

OLIGOCENE: One of the epochs of the Tertiary period. It lasted from 40 million to about 26 million years ago.

ORDOVICIAN: A period of the Lower Palaeozoic era. Lasted from 500 million to 440 million years ago. (Named after the Ordovices, a tribe from north-central Wales.)

OROGENY: A period of mountain building, normally caused by a collision between two or more of the earth's crustal plates.

PALAEOCENE: The first epoch of the Tertiary period. Started 65 million years ago.

PALAEOZOIC: The "ancient life" era, lasting from about 570 million to 225 million years ago.

PEBIDIAN: A formation of volcanic rocks of Pre-Cambrian age. (Named after Pebidiog, the medieval name for St. David's peninsula.)

PERIOD: One of the main units of geological time. Each era is divided into a number of periods.

PERMIAN: The last period of the Palaeozoic era. It followed the Armorican Orogeny and lasted from 280 to 225 million years ago.

PILLOW LAVA: Lava laid down in the sea, so that it has the appearance of pillows piled one on top of another.

PLATFORM: A spectacular flat erosion surface, often cut by wave action across varied rock strata.

PLEISTOCENE: The main epoch of the Quaternary period, having started about 2 million years ago. The time of the most recent Ice Age.

PLIOCENE: The last epoch of the Tertiary period. It followed the Alpine Orogeny of Miocene times.

PRE-CAMBRIAN: The immensely long geological period lasting from the birth of the earth until 570 million years ago. Note that this period was very much longer than the rest of geological time put together.

PYROCLASTIC ROCK: Igneous rock formed as a result of violent volcanic explosions. May be composed of agglomerates and tuffs.

QUATERNARY: The most recent geological period. It is still with us.

QUARTZITE: A sedimentary or metamorphic rock made up largely of quartz grains or crystals.

RHYOLITE: A fine-grained grey or white lava, the volcanic equivalent of granite.

SANDSTONE: A sedimentary rock composed mainly of quartz grains and representing a compacted layer of sand.

SEDIMENTARY ROCK: Rock composed of organic or inorganic sediments, directly laid down after erosion and transport by the agents of denudation.

SHALE: A fine-grained sedimentary rock formed mainly from particles of silt and clay. It may be finely stratified.

SILURIAN: One of the periods of the Lower Palaeozoic era. (Named after the Silures, an ancient tribe from south-east Wales).

SLATE: A dense, fine-grained metamorphic rock which readily splits into thin, smooth plates.

STRATUM: A distinct bed or layer of rock. Plural: strata.

SYNCLINE: A trough or downfold in the rocks, usually produced as a result of pressure exerted from the sides.

SYSTEM: Term used for the sequence of strata laid down during a single geological period.

SOLIDIFICATION: The hardening of igneous materials into layers of solid rock.

TERTIARY: The "third life" period lasting from 65 million to 2 million years ago. Part of the Cenozoic era.

TRIASSIC: The first period of the Mesozoic era. Started about 225 million years ago.

TILL: A deposit laid down directly by glacier ice. Very variable in character. (Referred to in the older literature as "boulder clay".)

TUFF: Rock composed of fine-grained materials (often ash) ejected by a volcano. It may be arranged in beds which have accumulated under water.

UNCONFORMITY: A contact along which a younger group of rocks rests upon the eroded surface of an older group of rocks. Often the beds of the older group of rocks are truncated at the unconformity.

VOLCANIC ROCK: Rock laid down as a result of volcanic action. Examples are lavas, ashes, tuffs and agglomerates.

VESICULAR STRUCTURE: Structure in an igneous rock in which cavities (or vesicles) can still be seen. The cavities may be filled with later mineral crystals.

Some Useful References

Barrett, J. H., 1974, *The Pembrokeshire Coast Path* (Long-Distance Footpath Guide No. 3), HMSO, London, 118 pp.

Bassett, D. A. and Bassett, M. G. (editors), 1971, *Geological Excursions in South Wales and the Forest of Dean,* Geologists' Association, Cardiff, 267 pp.

Brown, E. H., 1960, *The Relief and Drainage of Wales,* University of Wales Press, Cardiff, 186 pp.

Evans, D. E., 1973, *Pembrokeshire Coast Scenery,* Nat. Museum of Wales, Cardiff, 36 pp.

Evans, R. O. and John, B. S., 1977, *The Pembrokeshire Landscape,* Five Arches Press, Tenby, 135 pp.

George, T. N., 1970, *South Wales* (British Regional Geology), HMSO, London, 152 pp.

John, B. S., 1972, *The Fishguard and Pembroke Area* (British Landscapes through Maps, No. 16), Geographical Association, Sheffield, 35 pp.

John, B. S., 1973, *The Rocks: Geology of Pembrokeshire,* Pembrokeshire Handbooks, Llanychaer, 32 pp.

John, B. S., 1976, *Pembrokeshire* (British Topographical Series), David and Charles, Newton Abbot, 240 pp.

John, B. S., 1976, *Scenery of Dyfed* (The Face of Wales Series, No. 3), Greencroft Books, Newport, Pembs., 28 pp.

Owen, T. R., 1973, *Geology Explained in South Wales,* David and Charles, Newton Abbot, 211 pp.

Sparks, B. W., 1971, *Rocks and Relief,* Longmans, London, 404 pp.

Steers, J. A., 1964, *The Coastline of England and Wales,* Cambridge University Press.